A Delicate Balance

Portfolio Analysis and Management
for Intelligence Information
Dissemination Programs

Eric Landree, Richard Silberglitt, Brian G. Chow,
Lance Sherry, Michael S. Tseng

Prepared for the National Security Agency
Approved for public release; distribution unlimited

NATIONAL DEFENSE RESEARCH INSTITUTE

The research described in this report was prepared for the National Security Agency. The research was conducted in the RAND National Defense Research Institute, a federally funded research and development center sponsored by the Office of the Secretary of Defense, the Joint Staff, the Unified Combatant Commands, the Department of the Navy, the Marine Corps, the defense agencies, and the defense Intelligence Community under Contract W74V8H-06-C-0002.

Library of Congress Cataloging-in-Publication Data is available for this publication.

ISBN 978-0-8330-4909-4

The RAND Corporation is a nonprofit research organization providing objective analysis and effective solutions that address the challenges facing the public and private sectors around the world. RAND's publications do not necessarily reflect the opinions of its research clients and sponsors.

RAND® is a registered trademark.

Published 2009 by the RAND Corporation
1776 Main Street, P.O. Box 2138, Santa Monica, CA 90407-2138
1200 South Hayes Street, Arlington, VA 22202-5050
4570 Fifth Avenue, Suite 600, Pittsburgh, PA 15213-2665
RAND URL: http://www.rand.org/
To order RAND documents or to obtain additional information, contact
Distribution Services: Telephone: (310) 451-7002;
Fax: (310) 451-6915; Email: order@rand.org

Preface

This publication describes the application of the RAND Corporation's Portfolio Analysis and Management Method (PortMan) to the evaluation of the National Security Agency's (NSA) information dissemination program portfolio, which is managed by the NSA Information Sharing Services (ISS) division. RAND's PortMan method enables the data-driven analysis of project portfolios and provides a means to monitor the progress of potentially high-value projects. It also allows portfolio managers to monitor the impact of any mitigation strategies they undertake, ensuring that the portfolio's highest potential value is achieved. For this project, RAND researchers first employed the Delphi method, a process for eliciting group opinion by a series of questionnaires with selective feedback from earlier responses, to collect expert opinion from the ISS Senior Leadership Group. This allowed for an estimation of value and risk for each project. RAND then used these estimates, together with cost information provided by ISS, to develop project rankings and to estimate the expected value-to-cost ratio for each project. RAND selected portfolios of projects that maximized the total expected value for the available program budget using a linear programming method and compared these results to ISS management's funding priorities.

This publication should be of interest to program and portfolio managers throughout NSA and the Intelligence Community, as well as project and program managers interested in or involved with information dissemination throughout government and industry. This unclassified report does not include detailed budget information.

This research was sponsored by NSA ISS and conducted within the Intelligence Policy Center of the RAND National Defense Research Institute, a federally funded research and development center sponsored by the Office of the Secretary of Defense, the Joint Staff, the Unified Combatant Commands, the Navy, the Marine Corps, the defense agencies, and the defense Intelligence Community.

For more information on RAND's Intelligence Policy Center, contact the Director, John Parachini. He can be reached by email at John_Parachini@rand.org; by phone at 703-413-1100, extension 5579; or by mail at the RAND Corporation, 1200 South Hayes Street, Arlington, Virginia 22202-5050. More information about RAND is available at www.rand.org.

Contents

Figures

Tables

Summary

This publication describes the application of RAND's PortMan for NSA's ISS division. PortMan enables data-driven analysis of project portfolios and provides a means for monitoring the progress of potentially high-value projects and associated risk-mitigation strategies, to ensure that this value is achieved.

In 2006, RAND performed a pilot study of the applicability of PortMan to ISS's research and development (R&D) project portfolio. The results of that study demonstrated that project rankings using PortMan, which were based on explicit value and risk metrics elicited from ISS management, were significantly different from those obtained using ISS's then-current method, which was based on undocumented, implicit metrics. While a definitive assessment of the final outcome of the two different rankings was beyond the scope of the pilot study, the RAND PortMan method did produce for the sponsor open, auditable, and transparent data that could then be used by program managers and senior decisionmakers to support program-related decisions. As a result of these findings and the added decision support materials generated in the PortMan pilot, ISS sponsored the broader study described herein.

PortMan evaluations are based on estimates of Expected Value (EV) of each project, defined as

$$EV = \textit{Value of Successful Implementation} \times \textit{Probability of Successful Implementation}.$$

Value of Successful Implementation is a measure of the benefit if the project is successfully implemented.[1] *Probability of Successful Implementation* is a measure of the difficulty or risk associated with either implementing an R&D project or sustaining an operations and maintenance (O&M) project.

RAND developed two different sets of metrics for estimating EV, one set for R&D projects and one set for O&M projects, based on elicitations of the important components of value and risk from ISS staff and analysis of documents provided by ISS management. To estimate the Value of Successful Implementation (i.e., value) and the Probability of Successful Implementation (i.e., risk) for each project, RAND con-

---

[1] *Value of Unsuccessful Implementation* is defined as zero.

xii   A Delicate Balance

ducted a Delphi consensus-building exercise using subject matter experts from ISS's Senior Leadership Group (SLG). The 17 projects included in the evaluation are listed and briefly described in Appendix A, and the questions presented to the ISS SLG during the Delphi exercise to estimate value and risk are included in Appendixes B and C. Appendix D includes an analysis of the Delphi exercise by project and by question. A high level of consensus among the SLG was reached after four rounds; for only four of 85 questions (five questions for each of 17 projects) were less than five of the ten SLG members in agreement on a single answer.

Figure S.1 is a plot of value versus risk for all 17 ISS projects listed in Appendix A, with O&M projects shown in orange and R&D projects in blue. Here value, which is plotted along the y-axis, is defined as the product of the four value components derived from the Delphi exercise using a 1-to-4 scale for the answers to the four value metric questions in Appendixes B and C.[2] The risk metric is defined as the answer to the risk metric question in Appendixes B and C. The risk scale is defined such that 1 corresponds to the answer *Substantial* and 4 to the answer *None*. In Figure S.1, the component of risk decreases as one moves from left to right along the x-axis. The size of the dot represents the level of consensus for each project: The smaller the dot the better the consensus. The gray lines show the standard deviation of the Delphi responses at the conclusion of the Delphi exercise. The EV of each project, calculated as *value* X *risk metric*, is shown in parentheses next to each dot. The four colored lines are constant EV contours at 5 percent (EV=51.2), 10 percent (EV=102.4), 15 percent (EV=153.6), and 20 percent (EV=204.8) of the maximum possible EV of 1,024.

Figure S.1 shows that the projects with the highest EV (SERV1), as well as the fifth-ranked project according to EV (TOOL1), have the highest risk compared to the rest of the portfolio. Thus, one clear recommendation that can be concluded from the PortMan analysis is to focus resources on risk-mitigation strategies or new R&D programs to support or replace these two projects. Figure S.1 also allows ranking of the projects according to EV, with the six projects falling below the 5-percent line identified as candidates for reduction or elimination.

However, PortMan also allows inclusion of project cost in order to balance value, risk, and cost in the analysis. In this case, ISS management provided RAND with the fiscal year 2008 (FY08) cost for each project and the total FY08 program budget. A linear programming (LP) model was used to select (from the 17 projects listed in Appendix A) a portfolio of projects that delivers the maximum portfolio EV (defined as the sum of the individual project EV for each project selected) for the available budget. Because projects have varying EV-to-cost ratios, this maximum EV portfolio includes three projects with less than 5 percent of the maximum EV (SUPP6, SERV4, and SUPP4) and excludes 2 projects with between 5 and 10 percent of the maximum

---

[2]   This method of combining the value metrics highlights the differences between projects. The scale assigns 1 to the answer *No*, 2 to *Very Little*, 3 to *Significantly*, and 4 to *Substantially*.

Figure S.1
Calculated Project Value Versus Risk

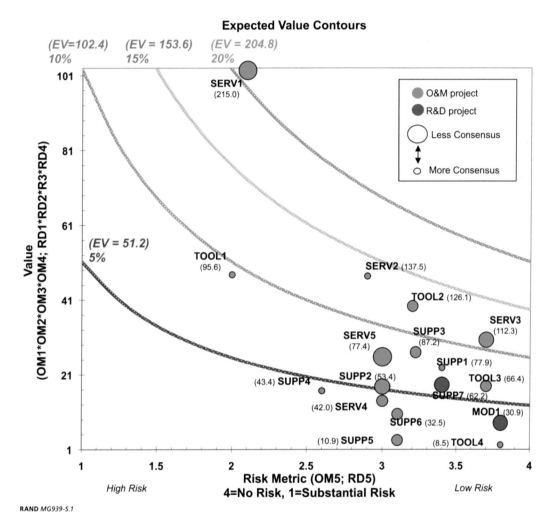

EV (SUPP3 and SUPP7), as well as one with greater than 10 percent of the maximum EV (SERV3).

Taking into account the fact that ISS partially funded some projects, the portfolio selected as achieving the maximum EV for the available budget was in good agreement with ISS's funding priorities for O&M projects.[3]

This PortMan analysis proved useful to ISS management in a number of ways. First, it generated reproducible and auditable data to support programmatic decision-making. Second, the Delphi exercise provided the ISS SLG with a venue in which to

---

[3]   If ISS funded a project at greater than or equal to 50 percent of the proposed cost, it was considered equivalent to being selected for the PortMan portfolio. See Chapter Three for details.

identify areas of consensus and non-consensus and to debate the latter. Finally, it provided data and analysis of EV versus program budget[4] and EV-to-cost ratios of individual projects that can be used by program managers and directors in discussions with supervisors and senior management, illustrated schematically in Figure S.2.

Figure S.2
Function of Portfolio Management for NSA ISS

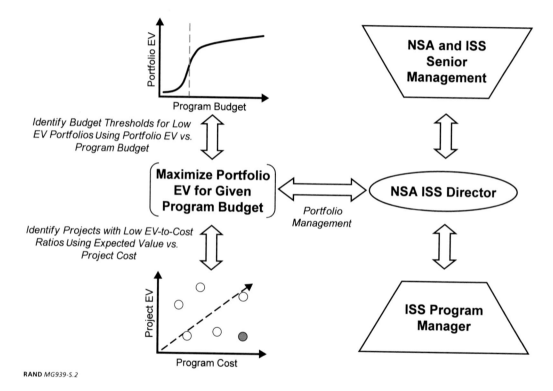

RAND *MG939-S.2*

---

[4] In addition to determining the maximum EV portfolio for ISS's FY08 budget, RAND used the PortMan with LP to determine maximum EV portfolios for budgets ranging from those sufficient to fund one project to those sufficient to fund all 17. See Chapter Three for details.

RAND drew the following conclusions from the results described in this report:

1. The RAND PortMan is a useful management method for both R&D and O&M portfolios.
2. Individual project cost can play an important role in achieving the highest expected value for a given portfolio.
3. RAND PortMan with LP is flexible enough that it may be applied to a single fiscal year or used to make strategic decisions that have implications for future fiscal years.
4. The Delphi method, as part of the portfolio management process, provides not only a mechanism for generating consensus, but also a forum for senior management to address and discuss areas of disagreement.

Acknowledgments

The authors would like to thank the individuals within the NSA ISS, as well as those throughout NSA, who supported us and provided their valuable time and energies toward making this study a success. In particular, we give special thanks to Marc Austin, Tawna Minton, Kelly L. Prescott, and the members of the NSA ISS Senior Leadership Group, who provided their ideas and their open and frank opinions and who invested significant time and effort to help us improve our method. We would especially like to thank Shannon R. Morris, without whose assistance and support this research would have been considerably more difficult and would have taken significantly longer to complete.

We are also indebted to our sponsor, the NSA ISS, for providing the support that allowed us to conduct this study.

We would also like to thank RAND colleague John Parachini for his continued support of our research efforts, as well as for the thorough and insightful comments that he provided throughout the course of this study.

Abbreviations

EV	Expected Value
FTE	full-time equivalent
FY08	Fiscal Year 2008
FY08C	Fiscal Year 2008 Case
FY12C	Fiscal Year 2012 Case
GAO	Government Accountability Office
ISS	Information Sharing Services
LP	linear programming
MOD	modernization activities
NSA	National Security Agency
O&M	operations and maintenance
OM1	Operations and Maintenance Metrics Question 1
OM2	Operations and Maintenance Metrics Question 2
OM3	Operations and Maintenance Metrics Question 3
OM4	Operations and Maintenance Metrics Question 4
OM5	Operations and Maintenance Metrics Question 5
PortMan	RAND's Portfolio Analysis and Management Method
R&D	research and development
RD1	Research and Development Metrics Question 1
RD2	Research and Development Metrics Question 2

RD3	Research and Development Metrics Question 3
RD4	Research and Development Metrics Question 4
RD5	Research and Development Metrics Question 5
SERV	network or Web-enabled services
SIGINT	signals intelligence
SLG	Senior Leadership Group (ISS)
SUPP	support
TOOL	hardware and/or software tools

CHAPTER ONE

Introduction: The Basics of Portfolio Management

This publication describes the application of the RAND Corporation's Portfolio Analysis and Management Method (PortMan) to the evaluation of the National Security Agency's (NSA) information dissemination program portfolio, which is managed by the NSA Information Sharing Services (ISS) division. As we focus herein almost exclusively on portfolio management, a brief description of the field will help provide a backdrop for our discussions. While there are many sources published in the literature and presented at conferences that delve deeply into this field (e.g., see Cooper, Edgett, and Kleinschmidt, 1998; Maizlish, Handler, and Nelson, 2005; Silberglitt and Sherry, 2002; and presentations at The Corporate Portfolio Management Conference[1]), the short definition provided below should be sufficient to set the context of this report.

For our purposes, *portfolio management* may be defined as a means for assessing the contributions and balance of a collection of projects aimed at achieving a common goal. This is in contrast to assessing individual projects independently and against their own unique set of goals. While specific metrics for measuring contributions and assessing balance may vary, they will typically fall into one of three general categories: value, risk, or cost. Moreover, our use of the term *portfolio management* does not refer to any one particular methodology; rather it refers to a general approach for which many alternative methods exist. A non-exhaustive list of the types of methods that could be useful for conducting portfolio management might include, e.g., Balanced Scorecard (Nair, 2004), Applied Information Economics (Hubbard, 2007), IBM's Rational Method (Hanford, 2006), Earned Value Management (Fleming and Koppelman, 2005), as well as RAND's PortMan[2] method (Silberglitt et al., 2004).

Portfolio management has been used by industry to maximize potential return on investment (McKenna, 2006) or to identify possible emerging market opportunities (Adams et al., 2000/2001). Within government, portfolio management methods are also used to help maximize value for the investment of taxpayer dollars (GAO, 2007). In addition, most portfolio management methods have the added benefit of creating

---

[1] For more information see Institute for International Research, undated.

[2] *PortMan* is derived from "Portfolio Analysis and Management" and is defined and used in two other RAND publications that demonstrate the same method (Silberglitt et al., 2004; Chow et al., 2009).

records and data, which helps to render decisionmaking processes transparent and enables audits for fiscal responsibility and accountability.

The objectives of RAND's PortMan portfolio analysis and management method are to enable data-driven analysis of project portfolios and to provide a means for monitoring the progress of potentially high-value projects and, to ensure this value is achieved, associated risk-mitigation strategies. The ultimate benefit is decisionmaking that is informed by the latest data and analysis concerning value, technical performance, risk, and risk-mitigation strategies, in the best case absent personality-driven biases.

Applications of Portfolio Management for the National Security Agency Information Sharing Services Division

The function of the NSA ISS, the sponsor of the study described here, is to disseminate information to key NSA stakeholders and customers. Consequently, the ISS portfolio includes a diverse set of operations and maintenance (O&M) and research and development (R&D) projects aimed at supporting signals intelligence (SIGINT) development and dissemination. Given the diverse range of projects for which ISS is responsible, portfolio management methods can be a useful tool for identifying and funding projects that provide the best value in enabling national security SIGINT products to reach critical customers and stakeholders.

In 2006, the NSA SIGINT Bridge Office arranged a pilot study to gauge the applicability of RAND's PortMan method to ISS's R&D project portfolio. This RAND pilot study was conducted in parallel with ISS's existing project prioritization method, the 100 Coins exercise, in which each participant was given 100 units to distribute among the projects in any manner that they saw fit. In this exercise, the final project ranking was then based on the number of coins that each project received. Using the 100 Coins method, the projects were evaluated by each reviewer using their own internal implicit set of standards, without any formal record of how these standards were applied. Conversely, the RAND PortMan method used explicit questions and scales based on elements of value elicited from ISS management. The value and risk questions were presented to each participant, and their responses to these questions for each project were obtained and recorded simultaneously with their assignments of coins to each project. RAND then used the scales to rank the projects according to Expected Value (EV), based on the value components elicited from ISS management.

At the end of the pilot study, the value-based project rankings from the RAND PortMan method were compared to those from the 100 Coins method. The two methods produced markedly different project prioritizations, as illustrated in Figure 1.1. Green arrows in the figure denote projects whose coin rank was higher than their Port-Man EV rank, while red arrows denote projects whose coin rank was lower than their

Figure 1.1
Project Ranking Using 100 Coins and RAND PortMan Methods

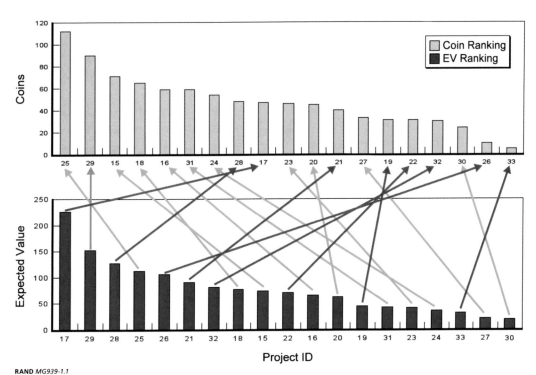

RAND *MG939-1.1*

PortMan EV rank. Only project 29 was ranked the same under both criteria. Of the top five projects as ranked by 100 Coins, only two were ranked in the top five using PortMan EV. When comparing the results of the two methods, as shown in Figure 1.1, there appears to be a systematic shift in the ranking of projects between the two methods. Projects that rank higher using PortMan end up having a significantly lower ranking using 100 Coins, and vice versa. These differences are especially important because ISS operates in a resource-constrained environment.

In addition to its difference from the 100 Coins rankings, the RAND PortMan method appealed to ISS management because it provides a transparent, open, and auditable method for estimating the value and risk of each project, along with the necessary data to support decisionmaking. Based on the results of the pilot study, ISS asked RAND to return the following year to perform a complete PortMan analysis of their portfolio that included both R&D and O&M projects and that allowed for a multi-round Delphi exercise using subject matter experts to develop consensus estimates of value and risk. This publication describes the results of that analysis.

Organization of This Monograph

Chapter Two provides descriptions of the RAND PortMan method, the R&D and O&M metrics, and how the Delphi method was used to collect expert opinion to estimate value and risk for each project. Chapter Two also includes an explanation of how individual project cost and available program budget influence the selection of project portfolios within the PortMan method.

Chapter Three presents the results of RAND's evaluation of NSA ISS's project portfolio, including estimates of value and risk and information on the expected value-to-cost ratio for each project. It also includes a set of portfolios that maximize total portfolio expected value for a range of program funding levels. These portfolios, as well as the project value-to-cost ratios, were constructed based on individual project cost estimates provided by ISS management.

Chapter Four offers observations and conclusions derived from the results presented in Chapter Three.

CHAPTER TWO

The RAND PortMan Method

Any portfolio management method should provide the capability to monitor project performance and to enable data-driven decisions to ensure that the highest potential value is achieved. The RAND PortMan method accomplishes this by estimating the components of value and risk for each project based on an agreed upon set of metrics that are linked to the functions or capabilities supported by this project and the entire portfolio. The components of value and risk are then used to analyze the risk versus reward of individual projects, as well as the balance across the entire portfolio. The results of this analysis may then be used to manage project performance, identify candidates for risk mitigation, and ensure balance and alignment with the overall objective of the portfolio.

Expected Value

The EV of an individual project is defined in PortMan as the product of the value if fully implemented and the probability of successful implementation:

$$EV = \textit{Value of Successful Implementation} \times \textit{Probability of Successful Implementation}.$$

Depending on the desired functions or capabilities being supported by the portfolio, there may be multiple ways in which a project can contribute Value of Successful Implementation.[1] In some cases it may be necessary to construct several value metrics, one for each different value component. The sum or product of each value metric may then be used to estimate the total Value for a particular project.[2] The use of a linear or

---

[1] The *Value of Unsuccessful Implementation* is defined as zero. Therefore, for the remainder of this report, the *Value of Successful Implementation* will be referred to simply as *Value*.

[2] There are many approaches that one may use for assessing individual project Value when there are multiple value components. It is possible to use any combination of mathematical operations upon the value metrics for each component to arrive at a final estimate. However, the relative weighing of these components must be taken into account. In another application of the PortMan method recently performed by RAND for the U.S. Army, the components are treated individually, and each component is assigned its own minimum threshold. Projects

nonlinear function for combining the different value components can have an effect on the overall ranking of the individual projects within the portfolio. This needs to be taken into account when designing the appropriate value metric. For this study, the product of the individual value components was used.[3]

Probability of Successful Implementation refers to the difficulty or risk associated with implementing (for R&D) or sustaining (for O&M) a given project. While the metrics for assessing Value are linked to the functions or capabilities that the portfolio is trying to achieve, the metrics for risk depend on the type of project. For example, the risk of implementation for R&D projects is typically associated with the difficulties involved with transitioning new technologies or integrating new capabilities into existing systems. It may also reflect challenges associated with acceptance or adoption of new technologies by the intended user community. The risk of implementation for O&M projects, on the other hand, is usually related to challenges associated with supporting legacy systems or more mature technologies. For example, O&M projects may incur risk as a result of the dwindling number of manufacturers able to provide replacement parts or the lack of available personnel with the expertise to repair or maintain systems supported by the project. It is interesting to note that the O&M risk for legacy systems tends to increase with time, while the risk associated with R&D projects may decrease over time as the projects address outstanding issues and technologies mature.

Metrics

RAND developed two different sets of metrics for estimating EV, one set for R&D projects and one set for O&M projects. These metrics were based on elicitation of the important components of value and risk from ISS staff and on our analysis of documents provided by ISS management.

Research and Development Programs

RAND developed the R&D value metrics using a vision of the future SIGINT dissemination architecture that was developed by ISS and vetted by them throughout the Intelligence Community. This future SIGINT dissemination architecture improves on the current architecture in four principal areas: (1) streamlining the process of preparing, manipulating, and disseminating SIGINT products; (2) speeding the rate with

---

are then selected for the portfolio such that the overall portfolio meets or exceeds all of the assigned minimum value thresholds. In this case, the weighing of the different components of Value is handled by independently adjusting the minimum threshold for each value component (See Chow et al., 2009).

[3] The product of the individual value components was used because the value components as defined for this study measured different aspects of value that were not viewed as being additive. The product, as opposed to the sum, of the value components provided a more appropriate (hybrid) value metric that combined these four components (aspects) of value analogous to a geometric, as opposed to an arithmetic, mean.

which SIGINT products can be disseminated; (3) eliminating duplicative processes and services; and (4) enabling new services, tools, and capabilities to be seamlessly integrated into the architecture baseline. A separate value metric was constructed to reflect each of these areas of improvement.

The R&D risk metric estimates how difficult it would be to transition the findings, technologies, or capabilities resulting from the R&D project to the operational environment. Accordingly, this risk metric serves as the estimate of the Probability of Successful Implementation. Thus, the EV for each R&D project can be estimated as the product of this risk metric and the Value estimate derived by combining the R&D value metrics.

Operations and Maintenance Programs

It was determined that O&M projects contributed value to current ISS operations in four different areas: (1) contribution to current operations, (2) criticality to the users of products produced, (3) ability to provide margin or ancillary service in the event that other capabilities go offline, and (4) contribution to restoring full operations in the event of a disruption. Separate value metrics were constructed to assess the contribution of every O&M project to each of these areas.

The O&M risk metric estimates how difficult it is to support the current project. In the case of O&M, Probability of Successful Implementation corresponds to how difficult (or easy) it is to support the current project. Thus, the EV for each O&M project can be estimated as the product of this risk metric and the Value estimate derived by combining the O&M value metrics.

Value and Risk Estimation

There are two basic approaches for arriving at estimates of the value and risk metrics for each project. One approach is to have an individual analyst estimate the value and risk based on the best information available. The second approach uses a consensus-building method to help a group of subject matter experts estimate the risk and value metrics for each project. Both methods have been used in previous applications of the PortMan method (Silberglitt and Sherry, 2002; Silberglitt et al., 2004).

For this investigation, RAND was provided access to the NSA ISS Senior Leadership Group (SLG), which consists of program managers and directors who are familiar with the ISS portfolio. Given their availability, the RAND team chose to conduct a Delphi consensus-building exercise with this group of experts to estimate the value and risk for each project. The Delphi method is a systematic consensus-building process for obtaining expert opinion on a particular topic or question (see Helmer-Hirschberg, 1967). Recent studies have also investigated extending Delphi method as an electronic consensus-building tool (Wong, 2003; Gordon, 2007) and as a method for conducting foresight studies (Georghiou, 2001; Glenn and Gordon, 2003, 2007; Schwarz, 2006).

To employ the Delphi method in this setting, we had to develop questions for the members of the SLG to answer to determine the value and risk for each project. We also needed every member of the Delphi exercise to have access to the information they required to arrive at answers for each of the questions. Therefore, the RAND team prepared a short narrative for each project based on data derived from consultations with ISS staff and provided it to the members of the SLG during the Delphi exercise. In addition, we developed a spreadsheet tool to facilitate each round of the Delphi and to capture the anonymous responses by the SLG members to each question. This allowed the RAND team to monitor the consensus among the SLG members and to identify specific areas of disagreement.

The Delphi exercise employed in this study consisted of four separate rounds. The results for each round were aggregated, stripped of identifying information, and presented back to the SLG members in the subsequent round. After the third round, the areas of non-consensus were presented for discussion at a meeting of the SLG as part of a RAND-facilitated workshop. The final estimates of the value and risk metrics and the level of consensus for each metric were then presented, along with preliminary recommendations, to the Director of the ISS and members of the SLG.

Portfolio Analysis

Attempting to strategically manage a portfolio of projects to achieve the highest potential value for an organization without the ability to assess how well it is performing relative to its intended object(s) is difficult, if not impossible. In portfolio management, much like in navigation, it is important to start with a clear map of where one is in order to determine the optimal path (i.e., series of decisions) to reaching the desired goal. In the RAND PortMan method, the process of the Delphi participants discussing the value or risk metrics for a particular project provides the program manager with information about the status of their current program, both on an individual project level and from a portfolio perspective. Consensus is not necessarily achieved for all metrics. Those metrics for which consensus is not reached can inform the program manager as to what factors (e.g., lack of information on the part of the Delphi participants) are the source of the disagreement. It is then possible for program managers to both manage the balance of risk versus reward across the portfolio and to assess the relative progress toward (or way from) achieving the organization's overall objective.

Consensus

During each Delphi round, the RAND team monitored the level of consensus of the answers to each of the questions that determined the value and risk metrics for each project. The ISS portfolio consisted of 17 projects. The Delphi participants were asked five questions for each project (i.e., one question for each metric), for a total of 85 questions.

Ten members of the ISS SLG participated in the Delphi exercise. Consequently, the RAND team defined majority consensus as six members (i.e., one-half of the group plus one) or more agreeing on the answer to a question. Near-consensus was defined as five members of the Delphi group agreeing on the same answer, and non-consensus as less than half of the SLG or fewer agreeing on any one answer to a particular question. As will be demonstrated in detail in Chapter Three, a high level of consensus was achieved among the participants in the Delphi exercise. After four Delphi rounds, there were only four questions out of 85 for which there was not either majority consensus or near-consensus.

Risk Versus Reward

Once the value and risk metrics are estimated for each project, it is possible to assess the balance of risk and potential reward from a portfolio perspective. Figure 2.1 illustrates the relative regions of value and risk for a hypothetical portfolio. The axes are the value and risk metrics, with the value metric increasing along the y-axis and the

Figure 2.1
Sample Project Portfolio Showing Projects and Expected Value Contours

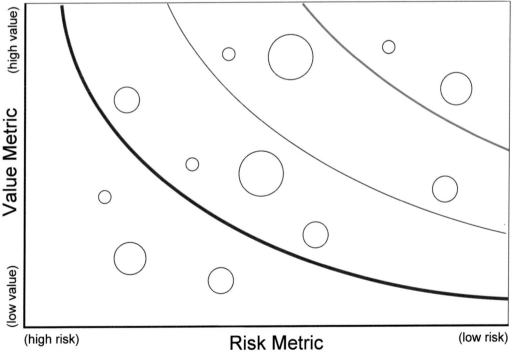

risk metric decreasing along the x-axis. For simplicity, in this example we assume only a single value metric. In Chapter Three we will combine the individual value metrics obtained from the SLG Delphi exercise to obtain an estimated value for each project. Each project is represented by a single circle or spot, the diameter of which reflects the consensus or agreement on the metrics for each project; the lower the consensus the larger the spot size, and the higher the consensus the smaller the spot size. The solid curves are contours of equal EV (i.e., value metric times risk metric).

Projects located in the upper right region of Figure 2.1 are the most desirable and have the highest value along with the lowest degree of risk. R&D projects in this region have the highest probability of transitioning to the operational environment and providing highly valued capabilities, while O&M projects in this region are the easiest to support and provide highly valued capabilities. Conversely, projects in the lower left quadrant are the least desirable, being of low value and high risk. Projects located in this region may be candidates for additional support (to increase one or more of the components of value) or potentially could be terminated to provide opportunity for other projects. Projects in the upper left quadrant represent those with high value and high risk. R&D projects in this quadrant require risk reduction or mitigation strategies. Ideally, project managers should pursue strategies that preserve the project value while reducing the component of risk, thus moving the project to the upper right quadrant of the figure. Projects in the lower right quadrant provide little value, but incur little risk.

Nominally, there is also a defined minimum acceptable EV that every project in the portfolio should exceed, represented in Figure 2.1 by the purple line. For R&D projects, such a minimum EV portfolio can be balanced in terms of the distribution of high-risk, high-value versus low-risk, low-value projects with similar EV.

The description of the various regions of Figure 2.1 for R&D projects presented above also applies to O&M projects. However, the appropriate level of balance for an O&M portfolio differs from that of an R&D portfolio: In general, an O&M portfolio that is closely coupled to mission-critical operations may have a much stronger aversion to supporting high-risk projects. Therefore, a balanced O&M portfolio may have proportionally more projects on the right-hand side of the graph. Alternatively, it may only be able to tolerate projects that exceed higher minimum required EV, such as the green line in Figure 2.1. However, there may be situations in which supporting a high-risk, high-value project is unavoidable, as is the case with many legacy systems. In addition, over time as projects are no longer supported, they will naturally move from right to left as the technologies and expertise to support them become increasingly difficult to secure. This is the opposite of what occurs with R&D projects, which tend to move from left to right as technologies mature. Therefore, O&M projects located in the upper left quadrant would indicate projects in need of risk mitigation, or possibly candidates for future R&D projects aimed at replacing legacy systems that have become difficult to support.

Cost Analysis

The ability to chart the relative risk and reward of projects using a common set of metrics is fundamental to successful portfolio management. It provides a graphical representation of where current of projects are and generally where they need to go in order to achieve both high value and low risk. However, graphs such as Figure 2.1 do not reveal the cost associated with trying to move projects to regions of lower risk and higher value. In a financially unconstrained environment, any amount of funds can be expended to achieve high-value, low-risk projects. However, in a fiscally constrained environment in which resources are scarce, the EV-to-cost ratio must be taken into account to assure that there is balance between value, risk, and cost.

The PortMan method accomplishes balance between value, risk, and cost using a linear programming (LP) model. The LP model uses information about individual project cost and the overall program budget to select a portfolio of projects that achieves the maximum portfolio EV for the available budget.[4] For this study, the LP model consisted of two components: (1) an objective function that we sought to maximize and (2) a set of constraints. The *objective function* was defined as the sum of the EV of every project selected for a particular portfolio[5], referred to as the *portfolio EV*. The constraint was that the *total cost* for the portfolio, defined as the summed cost of each individual project contained within the portfolio, must not exceed the available budget. In this study, each project in the portfolio was either fully funded or not funded (i.e., not selected as part of the portfolio). While the LP method is capable of including information concerning partially funded projects,[6] the information necessary to include partially funded programs was not available in this study.

The LP model is also capable of taking into account interdependencies between individual projects. For example, it is possible to construct a rule that says "only select project B if project A is also selected" to reflect the fact that project B may depend on some component of project A. Therefore, without project A, there is no value in funding project B. For this study however, ISS did not provide detailed information on the interdependencies that exist between projects. Thus, the simplifying assumptions described above were adopted for the analysis.

The fiscal year 2008 (FY08) cost for each individual project and the total FY08 program budget were provided to RAND by ISS. This cost information was integrated into the LP portfolio selection process and analyzed for two separate cases. The first

---

[4] For this study, the LP model was implemented using Microsoft® Excel.

[5] This definition assumes that the projects are independent and nonduplicative. It also sums the combined value metrics for each selected project rather than the individual value metric components.

[6] For example, projects that provide contractor support may decrease linearly in value with decreasing funds. Projects that rely on purchasing specific hardware or software may have a very high minimum cost threshold, below which the systems can no longer be afforded and the entire value goes away. Alternatively, projects that are primarily focused on sustaining or maintaining current systems may be able to adsorb significant decreases in budget before the value begins to decrease significantly.

case excluded project lifecycle cost. The goal of the LP objective function in this case was to spend all of the funds available for the current fiscal year. This was defined as the *Fiscal Year 2008 Case* (FY08C). The second case included theoretical lifecycle costs that the project would incur over the five-year defense planning cycle out to fiscal year 2012. This was defined as the *Fiscal Year 2008 to Fiscal Year 2012 Case* (FY12C). For FY12C, RAND assumed a fixed 3-percent annual inflation of project costs over the five-year period and a flat ISS budget. If a particular program could not be supported over the entire five-year period, it was not included in the final portfolio selection, even if there were sufficient funds to afford it in the current fiscal year. While RAND held these parameters fixed for this study, the LP method allows parameters such as the project cost inflation rate and program budget to be set independently for each project and for each fiscal year.

CHAPTER THREE

PortMan Evaluation of the NSA ISS Portfolio

Our evaluation of the NSA ISS division's portfolio involved a multistep process that occurred over a period of several months. The initial step involved analyzing the projects within ISS's current portfolio as well as the office operations to develop appropriate R&D and O&M metrics. After the metrics were completed, the Delphi exercise was organized to coincide with the ISS SLG's regularly scheduled meetings. The exercise required approximately four weeks to complete from start to finish. At the conclusion of the Delphi exercise, RAND conducted an analysis of the influence of project cost on portfolio selection. A high level of commitment and support from NSA ISS staff and management enabled the RAND team to conduct this PortMan evaluation at an efficient and deliberate pace.

ISS Project Descriptions

The ISS program portfolio that we evaluated consists of 17 projects. Five of the projects involve developing network or Web-enabled services (SERV). Seven of the projects are support activities (SUPP) that provide personnel or other resources to help process and create final SIGINT products and reports or to help maintain current systems. Four of the projects are developing specific hardware and/or software tools (TOOL) in support of ISS's primary function. These tools were characterized as standalone pieces of hardware or software intended to do a specific task, as opposed to services, which potentially involved several systems made up of multiple pieces of hardware and/or software to provide a capability or function. Lastly, one project involved integration and modernization activities (MOD) intended to improve and enhance the current baseline operations. A brief description of each of the ISS projects analyzed in this study is presented in Appendix A.[1]

---

[1] The project descriptions listed in Appendix A have been shortened and lack sensitive specific details that were included in the project descriptions provided to the ISS SLG for the Delphi exercise.

Questions for Evaluating Value and Risk Metrics

During the Delphi exercise, each member of the ISS SLG was give a series of questions and a ranking scale to provide answers. At the conclusion of the final Delphi round, we converted each of the responses to a numerical score and averaged to arrive at a final estimate for each metric. As described in Chapter Two, the product of the four value metrics was used to estimate the Value of the project. The average calculated from the respondents' answer to the risk metric question was used as the estimate for the Probability of Successful Implementation.

Research and Development Questions

Table 3.1 shows the five R&D questions that were developed to assess the value and risk for R&D projects in the ISS portfolio.

Each value metric question had four possible answers from which the Delphi participants could select (the corresponding numerical value that RAND used to estimate the metric is given in the parentheses to the right of each answer.): No (1), Very Little (2), Significantly (3), and Substantially (4). The risk metric question also had four possible answers: None (4), Small (3), Significant (2), and Substantial (1). The questions in Table 3.1 were provided to the ISS SLG members along with a description of the desired end state for each metric, specific example questions, and numerical thresholds (e.g., percentages) for each of the four possible answers, shown in Appendix B.

Operations and Maintenance Questions

Table 3.2 shows the five O&M questions that were developed to assess the value and risk for O&M projects in the ISS portfolio.

The possible answers and numerical values for each of the O&M value and risk metric questions were the same as those used for the R&D value and risk metric questions. The questions in Table 3.2 were provided to the ISS SLG members along with a description of the desired end state for each metric, specific example questions, and numerical thresholds (e.g., percentages) for each of the four possible answers, shown in Appendix C.

Table 3.1
R&D Value and Risk Metric Questions

Metric	Question
Value	RD1. Does it streamline the process?
Value	RD2. Does it speed up dissemination?
Value	RD3. Does it eliminate duplication?
Value	RD4. Does it improve scalability?
Risk	RD5. What is the implementation risk?

Table 3.2
O&M Value and Risk Metric Questions

Metric	Question
Value	OM1. Does it contribute to current operations?
Value	OM2. Is it critical to current operations?
Value	OM3. Does it reduce chance and degree of interruption in operations?
Value	OM4. Does it enhance quick and full recovery?
Risk	OM5. What is the implementation risk?

Consensus Analysis

During the RAND PortMan pilot study conducted at the NSA in 2006, the consensus after a single Delphi round exceeded 63 percent. In the pilot study, consensus was defined as more than half of the respondents to a single question agreeing on the same answer.[2]

In this investigation, four rounds of the Delphi process were conducted with all ten members of the SLG participating in all four rounds. The evolution of their consensus over the four rounds is shown in Figure 3.1. The numbers in each colored box correspond to the number of questions in that category. Consensus increased from approximately 52 percent in the first round to over 76 percent by the end of the fourth round. Near-consensus decreased from 27 percent in the first round to less than 19 percent, and non-consensus decreased from an initial value of approximately 21 percent (i.e., 18 questions) to less then 5 percent (i.e., 4 questions) in the final round.

Analysis of Value, Risk, and Portfolio Balance

The results from the 2006 RAND pilot study are presented in Figure 3.2.[3] Value and Risk Implementation[4] in Figure 3.2 were determined for each project using the same R&D metrics and questions described above in Table 3.1. The data collection included two different types of projects. The red dots represent projects for which funds were already committed, known as *must pays*. The green dots represent projects that were to be ranked in order of priority and considered for any available discretionary funds. The

---

[2] In the RAND pilot study there were a total of nine participants in the Delphi analysis. However, not all participants responded for all responses to a particular question, not on the total number of respondents. The minimum number of respondents for a given project in the pilot study was 4 and the maximum was 9.

[3] The RAND pilot study included only R&D projects; therefore only the R&D questions shown in Table 3.1 and Appendix B were used.

[4] *Risk Implementation* is the same as *Probability of Successful Implementation* or *risk* as defined in Chapter Two.

Figure 3.1
Consensus During Subsequent Rounds of Delphi Analysis

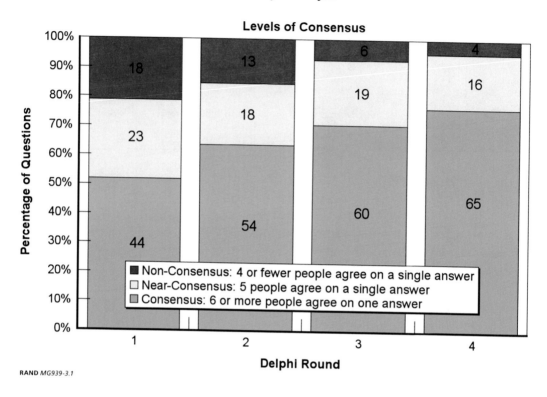

diameter of each project dot on the chart reflects the level of consensus: The smaller the diameter of the dot, the higher the consensus. Consensus was calculated based on the number of responses to each question. Gray lines denote uncertainty, as determined by the standard deviation of the responses. The red crosshairs represent average values for all must pay (red) projects, and green crosshairs represent average values for all discretionary (green) projects. Individual project descriptions were not provided to RAND or to the participants. Therefore, the participants needed to rely on their own knowledge of the project to answer each value or risk question. For the pilot study, individual projects were numbered from 1 to 33.

The must pay projects (red) show a bimodal distribution with a majority of the projects located in the lower right quadrant. The remaining must pay projects were located near the center of the chart, and appeared to have a significantly higher degree of value and risk compared to the rest of the must pay projects.

By contrast, the discretionary projects (green) had a more continuous distribution. However, it is important to note that the level of risk for all of the projects in the portfolio was low. According to the risk scale that was used (see Appendix B), the risk for all of the projects ranged from essentially no risk (i.e., "No known technical, per-

Figure 3.2
PortMan Expected Value Results for NSA ISS RAND Pilot Study

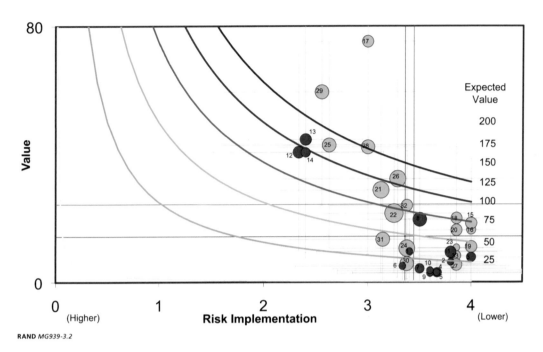

sonnel, resource, or process issues that may impede full implementation") to low risk (i.e., "Technical, personnel, resource, or process issues that may impede full implementation are difficult, but can be addressed using proven methods"). The results suggest that the projects included in this portfolio would not include enough risk to constitute a balanced R&D portfolio, even if all the projects were funded.

Between the RAND pilot study and the PortMan analysis that is the focus of this publication, most of the projects were reorganized or significantly changed. Therefore, there was no attempt on the part of the RAND team to compare the results of the pilot study with those of the latest study on a project-by-project basis.

An important difference between the pilot study and the PortMan analysis described here is the inclusion of both R&D and O&M projects in the latter. Figure 3.3 shows the Value[5] versus risk metric for both R&D (blue) and O&M (orange) projects. As before, the size of the dot that represents each project corresponds to the level of consensus for the five questions for that project, the smaller the dot the better the overall consensus. The gray lines show standard deviation of the responses at the conclusion of the Delphi exercise. The calculated total EV is shown in parentheses next to the title of each project (see Appendix A for project titles and descriptions). The four

---

[5] To reiterate: *Value* is determined as the product of the four component value metrics. This method of combining the value metrics highlights the differences between projects, as compared to the sum of the metrics.

Figure 3.3
Calculated Project Value Versus Risk

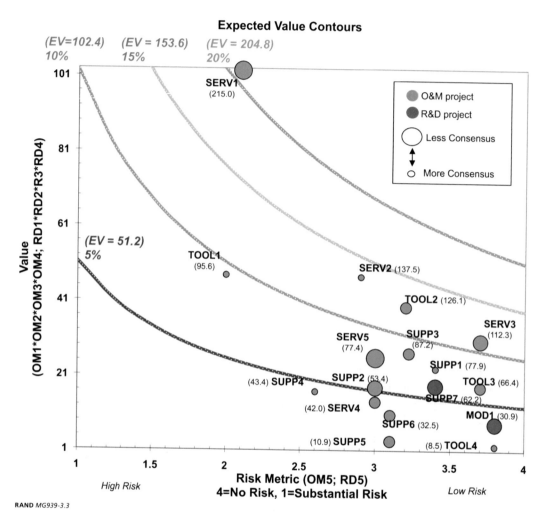

iso-EV lines correspond to 20 percent (EV=204.8), 15 percent (EV=153.6), 10 percent (EV=102.4), and 5 percent (EV=51.2) of the maximum possible expected value of 1,024. Out of all the projects in the portfolio, only one, SERV1, had a total EV that was greater than 20 percent of the maximum possible EV. The five highest-ranked projects based on expected value were SERV1 (214.99), SERV2 (137.53), TOOL2 (126.07), SERV3 (112.31), and TOOL1 (95.61). The bottom six projects, none of which achieved an EV of more than 5 percent of the maximum score, were SUPP4 (43.43), SERV4 (42.02), SUPP6 (32.54), MOD1 (30.87), SUPP5 (10.88), and TOOL4 (8.51).

According to these results, two projects that are among the top five according to EV also have the highest level of risk. Based on these initial results, one clear recommendation would be to focus resources on risk-mitigation strategies or new R&D programs to support or replace SERV1 and TOOL1. These EV results also suggest that in the event of a reduction in total budget, the projects below the 5-percent line should be preferentially cut or reduced. However, as we discuss in the next section, once information about cost and the available budget are included in the analysis, it is no longer the case that removing projects with the lowest expected value achieves the portfolio with the highest total EV.[6]

Figure 3.4 presents a comparison of the different components of value for each of the projects in the portfolio. The sum of value metrics OM1 and OM2 are plotted along the x-axis and the sum of metrics OM3 and OM4 (O&M projects are shown in orange) along the y-axis. The same plot also shows the sum of value metrics RD1 and

Figure 3.4
Comparison of Components of Value for O&M and R&D Projects

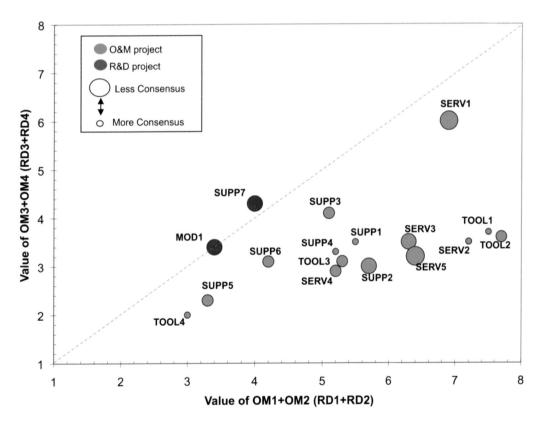

---

[6] For purposes of this investigation, R&D and O&M projects measured using different value and risk metrics were plotted and analyzed using the same numerical scale. This was done because the limited number of R&D projects in the portfolio made analyzing them separately from the O&M projects impractical. Throughout the subsequent analysis, the R&D and O&M projects and metrics are labeled and colored differently to distinguish one from the other.

RD2 compared to the sum of value metrics RD3 and RD4 (R&D projects are shown in blue). The dashed line defines the function y=x. Figure 3.4 shows that the O&M projects did not contribute as much to system recovery and additional margin (based on responses to questions OM3 and OM4, see Appendix C) as they did to current operations or criticality of the final products (based on responses to questions OM1 and OM2, see Appendix C). Another observation we can make from Figure 3.4 is that SERV1, the O&M project with the highest contribution to OM3 and OM4, also had the lowest consensus. For that project, only questions OM1 and OM2 had consensus; questions OM3 and OM5 (risk metric) had near-consensus, and question OM4 had non-consensus (see Appendix D, which shows consensus results by project and question). This suggests that, assuming all components of value are weighted equally, the O&M project portfolio is poorly balanced between contribution to operations and provision of margin and ability to recover from disruptions.

Cost Analysis

The RAND team used the cost information provided by ISS management together with the estimated project EV to create a scatter plot with cost along one axis and EV along the other axis. The points within this plot were used to calculate a linear least-squares best-fit line that was forced to go through zero. This line represented the notional EV/cost function defined by the current portfolio of projects. Projects located above the line provided relatively more value per dollar, while those projects below the line provided less value for the relative cost. Table 3.3 shows a list of the projects and indicates whether each project was above or below the portfolio EV/cost line.

Of the 17 projects in the portfolio, nine fell above the EV/cost line and eight below. Projects with a high EV did not necessarily have an EV-to-cost ratio that was better than the portfolio's average EV-to-cost ratio. For example, project SERV3 had the fourth-highest EV score (see Figure 3.3). However, the project's cost was sufficiently high that it had a below average EV-to-cost ratio. Another example is project SUPP3, which was among the top six projects according to EV score. However, this project had an EV-to-cost ratio below the portfolio's EV-to-cost ratio, suggesting that it is an expensive project for the value that it provides.

Portfolios for a range of different program budgets were then constructed using the LP model for both FY08C and FY12C. Figure 3.5 illustrates the portfolio selection for one of the two cases. Each column represents a different total program budget, and each cell is colored red or green depending on whether the project in that column was included in the portfolio for that particular program budget. Red indicates that the project was not selected for the portfolio, and green indicates that it was selected. The far right column shows the smallest budget, which contains sufficient funding for only a single project, SUPP1. Each subsequent column to the left represents an arbitrary

Table 3.3
Project Expected Value Versus Cost

Project Name	EV/Cost
SERV1	+
SERV2	+
SERV3	-
SERV4	+
SERV5	+
SUPP1	+
SUPP2	+
SUPP3	-
SUPP4	-
SUPP5	-
SUPP6	-
TOOL1	+
TOOL2	+
TOOL3	+
TOOL4	-
SUPP7	-
MOD1	-

NOTE: + = above the EV/cost line; - = below the EV/cost line.

increase in the total program funding. The column on the far left represents the largest budget, which contains sufficient funds to afford every project in the portfolio. The total portfolio EV, which is the sum of the individual EVs for each project selected for that portfolio, is given in the bottom row.

Note that there are some projects that are not selected for one program budget but that are included in the portfolio for a lower total program budget. For example, project TOOL4 is not selected in column Q. However, it is then included in a portfolio with a lower total program budget, shown in column P. This occurs as a result of the different EV-to-cost ratios for each project and LP method, which results in the selection of projects that combine to achieve the highest portfolio EV for the available budget. In some circumstances, moving from a higher to a lower total program budget means that an expensive project is no longer cost effective, and the funds that are freed

Figure 3.5
Portfolio Composition Based on Available Budget (FY12C)

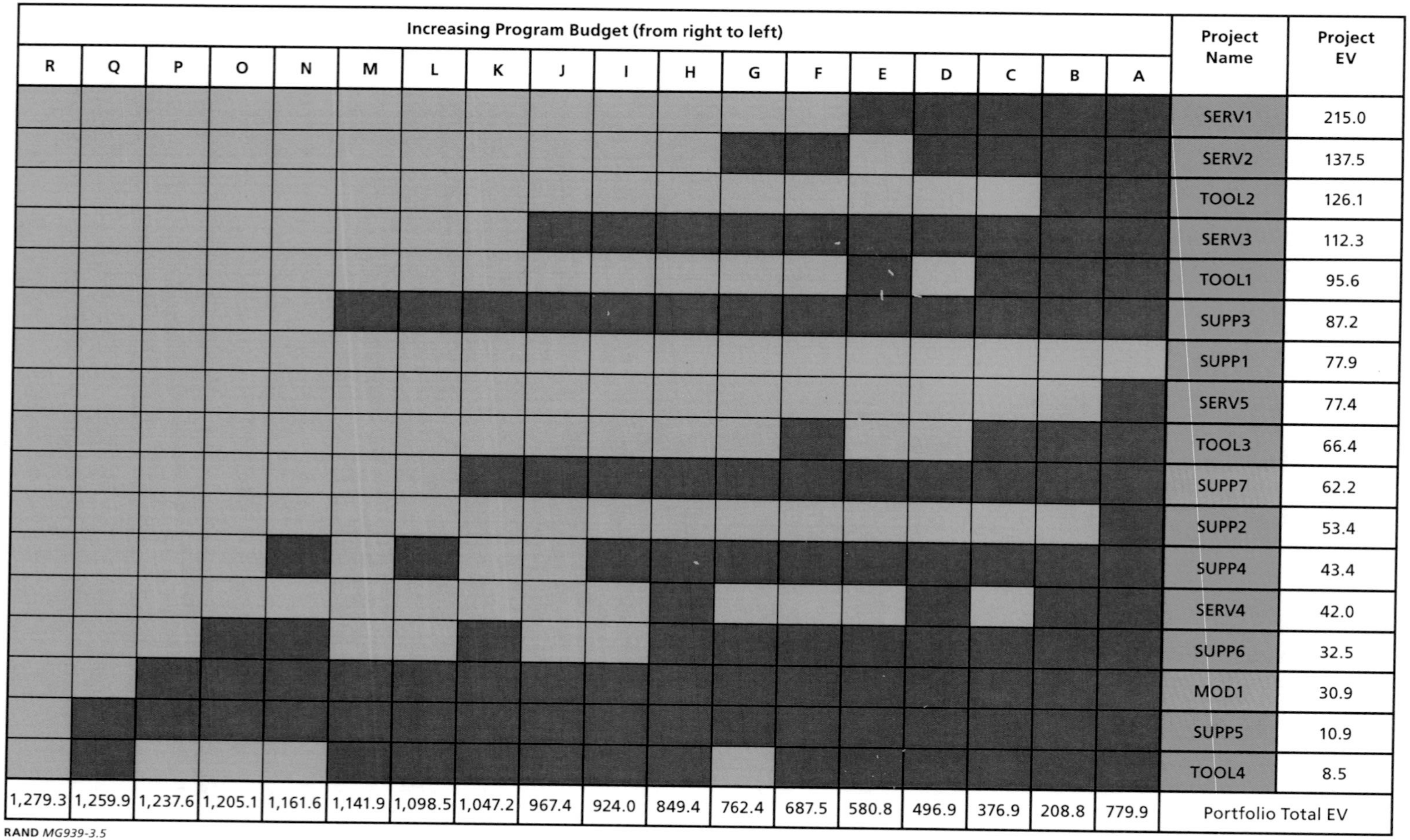

Increasing Program Budget (from right to left)

Project Name	Project EV	A	B	C	D	E	F	G	H	I	J	K	L	M	N	O	P	Q	R
SERV1	215.0																		
SERV2	137.5																		
TOOL2	126.1																		
SERV3	112.3																		
TOOL1	95.6																		
SUPP3	87.2																		
SUPP1	77.9																		
SERV5	77.4																		
TOOL3	66.4																		
SUPP7	62.2																		
SUPP2	53.4																		
SUPP4	43.4																		
SERV4	42.0																		
SUPP6	32.5																		
MOD1	30.9																		
SUPP5	10.9																		
TOOL4	8.5																		
Portfolio Total EV		779.9	208.8	376.9	496.9	580.8	687.5	762.4	849.4	924.0	967.4	1,047.2	1,098.5	1,141.9	1,161.6	1,205.1	1,237.6	1,259.9	1,279.3

RAND *MG939-3.5*

up may be used to select a less-expensive project, even if that project has a lower project EV. In all cases, the total portfolio EV increases as the total program budget increases.

Constructing portfolios for a range of total program budgets is useful for analyzing how decreases or increases in total program budget may influence both total portfolio value (i.e., Portfolio EV) as well as the portfolio's ability to provide or sustain certain capabilities. For example, as we can see in Figure 3.4, project SERV1 provides almost twice as much value along metrics OM3 and OM4 compared to all the other O&M projects. However, for program budgets less than that of Column F, there are insufficient funds to include SERV1 and still achieve the highest possible Portfolio EV. This suggests that, below this funding level, capabilities associated with those two metrics will decrease significantly unless a portfolio with less than the maximum total EV is chosen. In other words, in this program budget range, there is a tradeoff between total EV and value metrics OM3 and OM4, which represent the capabilities of margin and recovery.

The last two columns in Figure 3.6 provide a direct comparison of the ISS funding priorities to the results from RAND's PortMan analysis assuming no lifecycle costs (FY08C). The up-arrows indicate that the ISS funded the project at greater than 50 percent of the total proposed cost, and the down-arrows indicate that the project was funded at less than 50 percent of its total cost. As described in Chapter Two, because RAND was not provided with the effect of partial funding on project value, our portfolios fund each project either at 100 percent (selected) or not at all (not selected). A comparison of the two columns reveals strong agreement between the priority funding for the ISS office and the results returned from the RAND PortMan method. Nearly every O&M project that the RAND PortMan method selected for the portfolio was also funded at greater than 50 percent by the ISS division; the only exception was project SUPP6. However, the ISS division funded both R&D projects at greater than 50 percent, while the PortMan method recommended not funding either project. This lack of agreement concerning the R&D projects may be a result of the facts that (1) there were only two R&D projects and (2) they were compared to the O&M projects in the same portfolio.[7] Nonetheless, better than 82-percent agreement was achieved for the entire portfolio, and considering only the O&M projects, greater than 92-percent agreement was achieved.[8] This high level of agreement suggests that the

---

[7] As noted earlier, we plotted and analyzed R&D and O&M projects with the same numerical scale even though they were measured using different value and risk metrics because the limited number of R&D projects in the portfolio made analyzing them separately from the O&M projects impractical.

[8] For purposes of this analysis, if the ISS division funded a project at greater than or equal to 50 percent of the total proposed project cost, and the RAND PortMan method included the same project in the portfolio, they were considered to be in agreement. Similarly, if the ISS division funded a project at less than 50 percent, and the RAND PortMan method did not select the project for the portfolio, they were also considered to be in agreement.

**Figure 3.6
Comparison of PortMan Results Including Cost Information
(FY08C) with ISS Priorities**

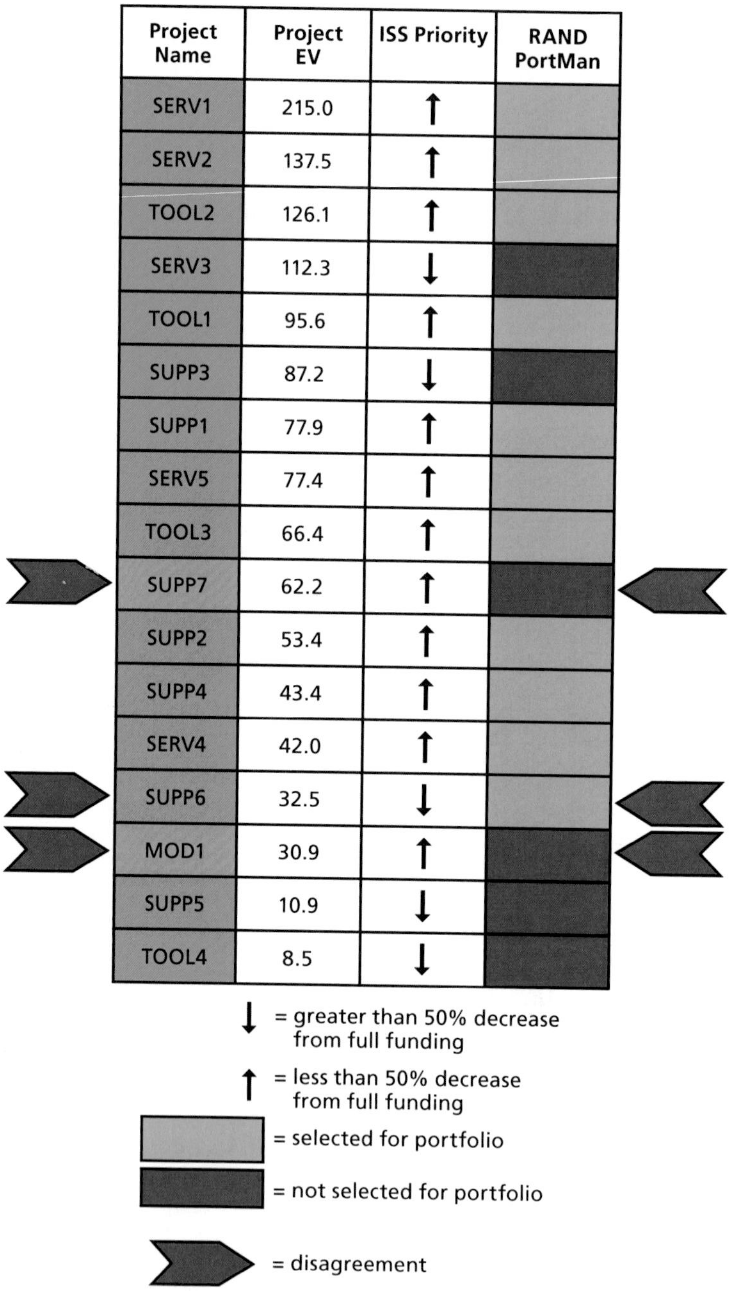

Project Name	Project EV	ISS Priority	RAND PortMan
SERV1	215.0	↑	
SERV2	137.5	↑	
TOOL2	126.1	↑	
SERV3	112.3	↓	
TOOL1	95.6	↑	
SUPP3	87.2	↓	
SUPP1	77.9	↑	
SERV5	77.4	↑	
TOOL3	66.4	↑	
SUPP7	62.2	↑	
SUPP2	53.4	↑	
SUPP4	43.4	↑	
SERV4	42.0	↑	
SUPP6	32.5	↓	
MOD1	30.9	↑	
SUPP5	10.9	↓	
TOOL4	8.5	↓	

↓ = greater than 50% decrease from full funding

↑ = less than 50% decrease from full funding

= selected for portfolio

= not selected for portfolio

= disagreement

O&M metrics, the estimated project EV scores, and the LP method used to assemble the portfolios were appropriate.

The RAND PortMan method, while demonstrating that ISS management's funding priorities are consistent with value and risk metrics, provided a number of additional benefits. First, it generated reproducible and auditable data that can be used to support programmatic decisionmaking. Second, the Delphi exercise provided the ISS SLG with a venue in which to identify areas of consensus and non-consensus and to debate the latter. Finally, it provided data and analysis of EV versus program budget and EV-to-cost ratios of individual projects that can be used by program managers and directors in discussions with supervisors and senior management. These data and analyses should prove valuable for program budget justifications, as well as for identifying and discussing with program managers projects that are not achieving an acceptable value or projects that appear to be excessively risky. Thus, PortMan can provide the ISS management with the information needed to assess the overall program portfolio and, where necessary, take corrective actions to address imbalances therein.

CHAPTER FOUR

Conclusions

We draw a number of conclusions from the results described in previous chapters:

1. **The RAND PortMan is a useful management method for both R&D and O&M portfolios.** Our approach provided sufficient value and insight to NSA ISS management that they chose to replace the 100 Coins method with the PortMan portfolio analysis and management method. In addition, the ability to compare the PortMan results with the intuitive programmatic decisions made by the NSA ISS senior leadership helped to validate the accuracy of the R&D and O&M metrics, as well as the results from the SLG Delphi exercise.

2. **Individual project cost can play an important role in achieving the highest expected value for a given portfolio.** In the absence of available cost information, the project ranking shown in Figure 3.3 would suggest that in the event of a reduction in budget, the projects with the lowest project EV should be cut first. However, inclusion of the project cost information resulted in a different outcome. To achieve the highest possible portfolio EV for a particular program budget, the project EV-to-cost ratio must be taken into account. Therefore, although project-level EV is a useful metric for ranking the relative contribution of individual projects within a portfolio, decisionmaking based on project-level EV alone may not lead to the highest possible portfolio EV for the available budget.

3. **RAND PortMan with LP is flexible enough that it may be applied to a single fiscal year or used to make strategic decisions that have implications for future fiscal years.** More specifically, it allows users to customize parameters related to program funding, as well as project-related expenses such as increases in individual project costs due to inflation and other causes, in order to approximate the real-world conditions under which most organizations operate.

4. **The Delphi method, as part of the portfolio management process, provides not only a mechanism for generating consensus, but also a forum for senior management to address and discuss areas of disagreement.** The process of developing metrics collaboratively allowed program managers and senior management to offer critiques and suggestions as to what constituted value and risk, essentially providing everyone a common view of the direction of the portfolio. It enabled them to asses and debate the relative merits of a given project using a consistent and explicitly stated set of transparent metrics. The data generated during the Delphi process also allowed program managers and ISS management to advocate for or defend specific projects or project actions to upper management.

Ultimately, the RAND PortMan portfolio analysis and management method, as applied in this study, allowed the members of NSA ISS to conduct transparent, data-driven risk-reward analysis and decisionmaking that helped to achieve the optimal overall portfolio balance and to identify areas for risk mitigation.

Project Descriptions

SERV1	O&M	This program includes a collection service, serial number generation, preparation, validation, release, and dissemination tools for text-based SIGINT reports.
SERV2	O&M	Government and user-generated networked information services.
SERV3	O&M	Web-based information request service for SIGINT post-publication reporting.
SERV4	O&M	Summarization and dissemination service of critical SIGINT to specific global customers.
SERV5	O&M	SIGINT tailoring and follow-up support service.
SUPP1	O&M	Staff support for generation of multimedia SIGINT products.
SUPP2	O&M	Information sharing support to partners in support of specific intelligence community missions.
SUPP3	O&M	Production support including pre- and post-publication guidance and review.
SUPP4	O&M	SIGINT reporting and dissemination policy development and guidance.
SUPP5	O&M	Harmonization of business and mission processes.
SUPP6	O&M	System administration support for all operational systems residing in the laboratory environment.
TOOL1	O&M	Full text storage and retrieval tool for reports issued by members of the intelligence community and open source new services.
TOOL2	O&M	Report repository with integrated rapid search and retrieval capabilities for sensitive-series reports.

TOOL3   O&M   Dissemination and tailoring interface for the exchange of multi-media and other SIGINT.

TOOL4   O&M   Customized dissemination device hardware.

SUPP7   R&D   Staff support for research and development activities.

MOD1   R&D   Service integration and modernization activities.

Research and Development Questions

This appendix includes the value and risk metric questions for each R&D project that were posed to the participants in the Delphi exercise. It also includes other information provided to the participants, such as a description of the desired end state assuming that all of the capabilities have been achieved, and example questions. The participants were given four possible answers to each question, along with corresponding scale thresholds to assist them in selecting the most appropriate answer.

RD1. Does it streamline the process?	
Desired end state: To have a process for preparing and disseminating final products, as well as storing, scanning, searching and retrieving information that requires less human intervention and/or less time to complete.	
Example questions to consider: - Does <u>this project or the people supported by this project</u> reduce the number of individuals (FTEs) or hours necessary?	
No	Has negligible effect on the number of individuals or amount of time needed to complete the task (5% or less decrease)
Very Little	Reduces the number of individuals or time required by > 5% to 33%
Significantly	Reduces the number of individuals or time required by > 33% to 66%
Substantially	Reduces the number of individuals or time required by > 66%

RD2. Does it speed up dissemination?	
Desired end state: Enable more products to be disseminated per unit of time.	
Example questions to consider: - Does <u>this project or the people supported by this project</u> increase the number of final products that can be disseminated per unit of time?	
No	Has minimal effect on the number of final products that can be disseminated (33% or less increase)
Very Little	Increases the number of final products that can be disseminated by more than 33% and less than 200%
Significantly	Increases the number of final products that can be disseminated between 200% and 500%
Substantially	Increases the number of final products that can be disseminated by more than 500%

RD3. Does it eliminate duplication?	
Desired end state: To have a unified system capable of processing all incoming information, reports, and requests; searching, storing and retrieving information; and disseminating products via all dissemination routes.	
Example questions to consider: - Does <u>this project or the people supported by this project</u> reduce the number of systems, resources, or processes needed?	
No	Has negligible effect on the systems, resources, or processes needed (5% or less decrease)
Very Little	Reduces the number of systems, resources, or processes needed by > 5% to 33%
Significantly	Reduces the number of systems, resources, or processes needed by > 33% to 66%.
Substantially	Reduces the number of systems, resources, or processes needed by more than 66%

RD4. Does it improve scalability?	
Desired end state: To achieve an integrated architecture that can incorporate new types of information and data files; new methods for scanning, retrieving and processing information; new types of products; and can distribute through all required or new dissemination routes seamlessly without the need to create new or substantially modify existing architectures.	
Example questions to consider: - Does this project or the people supported by this project help enable new capabilities to be developed or added without the need for a new architecture? - Does this project or the people supported by this project show progress toward enabling new types of information to be incorporated or integrated into current capabilities without the need for creating new architectures?	
No	No effect on current baseline (i.e., entirely new architectures or systems required to enable capabilities)
Very Little	Major changes (hardware, software, processes) still required to enable new capabilities
Significantly	Only minor changes to hardware, software, and algorithms necessary to enable new capabilities
Substantially	Enables new capabilities to be created and be integrated into current operations w/o any significant changes to architectures

RD5. What is the implementation risk?	
Desired end state: To achieve full implementation and successfully obtain the desired streamlining, faster dissemination, elimination of duplication, and improved scalability.	
Example questions to consider: - Are there technical problems that will impede implementation? - Are there personnel or other resource issues? - Are there potential process issues?	
None	No known technical, personnel, resource, or process issues that may impede full implementation
Small	Technical, personnel, resource, or process issues that may impede full implementation are easy to address
Significant	Technical, personnel, resource, or process issues that may impede full implementation are difficult, but can be addressed using proven methods
Substantial	Technical, personnel, resource, or process issues that may impede full implementation are difficult and cannot be addressed using known methods

Operations and Maintenance Questions

This section includes the value and risk metric questions for each O&M project that were posed to the participants in the Delphi exercise. It also includes other information provided to the participants, such as a description of the desired capability assuming that the project is fully supportable and operating as intended, and example questions. The participants were provided with four possible answers to each question, along with corresponding scale thresholds to assist them in selecting the most appropriate answer.

OM1. Does it contribute to current operations?	
Description of capability: The project maintains the capability to rapidly produce and disseminate high-quality and high-quantity content, as well as store, scan, search and retrieve relevant reports and information.	
Example question to consider: - How does/would <u>the project</u> contribute to the current capability either in terms of quality, quantity, or speed? - How does/would <u>the people supported by this project</u> contribute to the current capability either in terms of quality, quantity, or speed?	
No	Without project/people, 5% or less degradation of quality, quantity, or speed
Very Little	Without project/people, greater than 5% and up to 33% degradation in one or more areas of quality, quantity, or speed
Significantly	Greater than 33% and up to 66% degradation in two or more areas of quality
Substantially	Quality, quantity, and speed would be degraded by nearly > 66%

OM2. Is it critical to current operations?			
Description of capability: The activity provides critical services (e.g., production, dissemination, storage, search) for responding to requests and meeting expected dissemination requirements.			
Example questions to consider: - What is the perceived value of the services provided by this project? - What is the perceived value of the services that are supported by the people affiliated with this project? - What is the perceived value of the products provided by the services enabled by this project or the people that are supported by this project?			
	No	Service is considered to be of little to no value	Does not contribute to any valuable products or services
	Very Little	Service is considered to be of little value	Contributes to product or service that is considered valuable
	Significantly	Service is considered to be a considerable value	Contributes to up to three valuable services or products
	Substantially	Service is considered to be of the highest value	Contributes to more than three different services or products considered valuable

OM3. Does it reduce chance and degree of interruption in operations?	
Description of capability: The project provides additional margin or overhead capabilities to help maintain the operational capabilities in the event that other systems should fail or go off-line.	
Example questions to consider: - Does/would <u>this project or the people supported by this project</u> provide robustness or redundancy to the current capability in the event that other systems fail? - Does/would <u>this project or the people supported by this project</u> provide back-up for current capabilities or ability to "pick up some of the slack," in event other systems or capabilities cease to function?	
No	Provides no additional margin or robustness to current operations
Very Little	Able to assume some of the capabilities (up to 30%) for one or more individual functions
Significantly	Able to fully (100%) assume all the functions and capabilities of a single system, or able to contribute partially (>30%) to multiple functions (2 or more)
Substantially	Able to fully (100%) assume the functions and capabilities of multiple systems (2 or greater)

OM4. Does it enhance quick and full recovery?		
Description of capability: The project allows for a quick and full resumption of current operations.		
Example questions to consider: - Should/would current operations be interrupted, how does/would <u>this project or the people supported by this project</u> contribute to the degree to which capabilities can be restored or reconstituted? - Should/would current operations be interrupted, how does/would <u>this project or the people supported by this project</u> contribute to speed with which capabilities can be reconstituted to restore current ops?		
No	Has no impact on the ability to restore or reconstitute operations	Has no impact on speed with which full operational capabilities can be restored
Very Little	Enables only limited (up to 33%) op capabilities to be reconstituted	Able to decrease the time to restore current op capabilities by up to 33%
Significantly	Enables most (more than 33% and up to 66%) op capabilities reconstituted	Able to decrease the time to restore current op capabilities by more than 33% and up to 66%
Substantially	Enables full operational capabilities to be restored or reconstituted	Able to provide near instantaneous reconstitution of full operational capabilities

OM5. What is the implementation risk?		
Description of capability: All tools, material, and personnel necessary to maintain current operations for the next 5 years are readily available.		
Example questions to consider: - What is the current and future risk (up to 5 years) associated with maintaining capability enabled by <u>this project or the people supported by this project</u>, over anticipated program lifetime? - Is it or will it be difficult to find parts and trained staff (experienced with the hardware and software used in the current system) to support this capability within the next 5 years?		
None	All necessary parts for ops and repair of equipment are readily available	Staff with the necessary skill sets to perform all O&M operations are readily available
Small	Limited problems locating hardware or parts for maintenance or repairs	Certain O&M operations such as repairs may take somewhat longer (up to 33% longer)
Significant	Considerable problems locating hardware or parts for maintenance or repairs	Some ops may take considerably longer and be more expensive to perform (> 33% to 66%)
Substantial	Compatible parts for operation or repair are not available	Unable to perform certain rare but critical operations including repairs

Delphi Exercise Results by Project and Question

The evolution of the consensus by project for each of the four rounds is shown in Figure D.1. Note that after four rounds, every question for every project had achieved the same or higher level of consensus with only two exceptions. Question OM3 of project SUPP2 and question OM4 of project SUPP6 both went from having consensus in round one to near-consensus by the end of round four.

Several questions within specific projects also showed no improvement in consensus through four rounds of the Delphi method. Questions OM4 and OM5 of project SERV1, question OM1 of SUPP2, and question OM2 of SUPP5 all had the same degree of consensus through all four rounds.[1] In addition, SERV5 showed no change in consensus for any of the individual questions through all four rounds of the Delphi method.

---

[1] Although the level of consensus for certain projects and questions may not have changed through consecutive rounds of the Delphi method, the actual answers selected by the members of the SLG may have changed. For example, if half of the participants all agreed on the same answer for one question, and then in the consecutive round, half of the group agreed on a different answer for that particular question, the question was still marked as having near-consensus even though the agreed upon answer changed from one round to the next. The same is true for questions that had consensus. If a majority of respondents gave the same answer in one round, and then a majority selected a different answer in the next round, it still showed that consensus was achieved for both rounds.

Figure D.1
Increases (↑) and Decreases (↓) in Consensus by Project and Question for Each Delphi Round

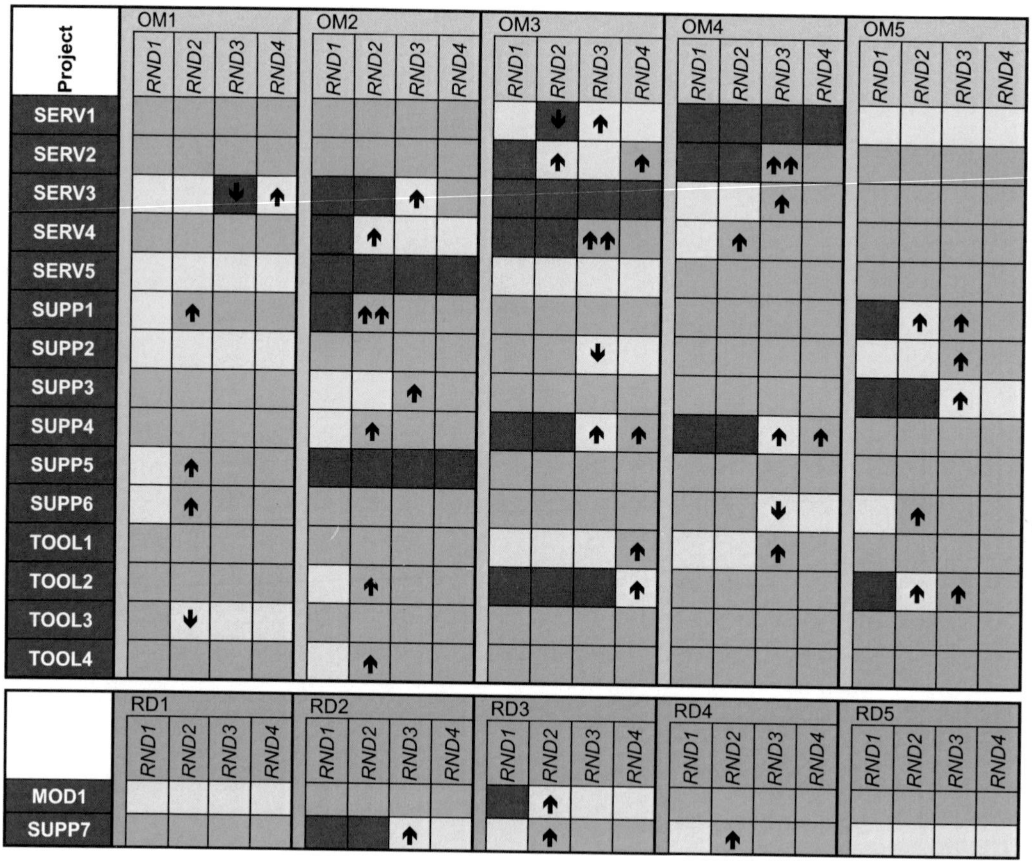

RAND *MG939-D.1*

Bibliography

Adams, Tom, Jeff Lund, Jack A. Albers, Michael Back, Jason McVean, and John I. Howell III, "Portfolio Management for Strategic Growth," *Oilfield Review*, Winter 2000/2001, pp. 10–19.

Chow, Brian G., Richard Silberglitt, and Scott Hiromoto, *Toward Affordable Systems*, Santa Monica, Calif.: RAND Corporation, MG-761-A, 2009. As of December 3, 2009: http://www.rand.org/pubs/monographs/MG761/

Cooper, Robert G., Scott J. Edgett, and Elko J. Kleinschmidt, *Portfolio Management for New Products*, Reading, Mass.: Addison-Wesley Publishing Company, May 1998.

Fleming, Quentin W., and Joel M. Koppelman, *Earned Value Project Management*, 3rd ed., Newtown Square, Pa.: Project Management Institute, 2005.

GAO—*See* Government Accountability Office.

Georghiou, Luke, "Third Generation Foresight—Integrating the Socio-Economic Dimension," *The Proceeding of International Conference on Technology Foresight*, Hirakawa-cho, Chiyoda-ku, Tokyo, March 2001. As of March 26, 2008: http://www.nistep.go.jp/achiev/ftx/eng/mat077e/html/mat077oe.html

Glenn, Jerome C., and Theodore J. Gordon, *Futures Research Methodology, v2.0*, Washington, D.C.: American Council for the United Nations University, 2003.

———, *2007 State of the Future*, 1st ed., New York, N.Y.: World Federation of UN Associations, 2007.

Gordon, Theodore, *Real Time Delphi*, World Federation of UN Associations, October 2007. As of March 26, 2008: http://mpcollab.org/MPbeta1/node/26

Government Accountability Office, *An Integrated Portfolio Management Approach to Weapon System Investment Could Improve DOD's Acquisition Outcomes*, Washington, D.C.: GAO-07-388, March 2007.

Hanford, Michael F., *Portfolio Management: Overview of New IBM Rational Methods*, March 15, 2006. As of March 20, 2008: http://www.ibm.com/developerworks/rational/library/mar06/hanford/

Helmer-Hirschberg, Olaf, *Systematic Use of Expert Opinions*, Santa Monica, Calif.: RAND Corporation, P-3721, 1967. As of December 3, 2009: http://www.rand.org/pubs/papers/P3721/

Hubbard, Douglas W., *How to Measure Anything: Finding the Value of "Intangibles" in Business*, Hoboken, N.J.: John Wiley & Sons, 2007.

Institute for International Research, *Portfolio Management for New Products and Services, February 2010*, Web page announcing conference, undated. As of October 26, 2009:
http://www.iirusa.com/portfolio/overview.xml

McKenna, Patrick, *Modern Portfolio Theory: Driving Project Portfolio Management with Investment Techniques*, August 15, 2006. As of March 20, 2008:
http://www.ibm.com/developerworks/rational/library/aug05/mckenna/index.html

Maizlish, Bryan, Robert Handler, and Ronald L. Nelson, *IT Portfolio Management Step-by-Step: Unlocking the Business Value of Technology*, Hoboken, N.J.: John Wiley & Sons, April 2005.

Nair, Mohan, *Essentials of Balanced Scorecard*, 1st ed., Hoboken, N.J.: John Wiley & Sons, 2004.

Schwarz, Jan O., "German Delphi on Corporate Foresight," Foresight Brief, No. 078, The European Foresight Monitoring Network, 2006. As of March 26, 2008:
http://www.efmn.eu/index.php?option=com_docman&task=doc_download&gid=78

Silberglitt, Richard, and Lance Sherry, *A Decision Framework for Prioritizing Industrial Materials Research and Development*, Santa Monica, Calif.: RAND Corporation, MR-1558-NREL, 2002. As of December 3, 2009:
http://www.rand.org/pubs/monograph_reports/MR1558/

Silberglitt, Richard, Lance Sherry, Carolyn Wong, Michael Tseng, Emile Ettedgui, Aaron Watts, and Geoffrey Stothard, *Portfolio Analysis and Management for Naval Research and Development*, Santa Monica, Calif.: RAND Corporation, MG-271-NAVY, 2004. As of December 3, 2009:
http://www.rand.org/pubs/monographs/MG271/

Wong, Carolyn, *How Will the e-Explosion Affect How We Do Research? Phase I: The E-DEL+I Proof-of-Concept Exercise*, Santa Monica, Calif.: RAND Corporation, DB-399-RC, 2003. As of December 3, 2009:
http://www.rand.org/pubs/documented_briefings/DB399/